Tall Tales

and

Mountain Trails

Stories and Poems © Mike Bryan
Illustrations © Skip Hanna
Book Design © Protégé Publishing
Published 2016

Contents

Biographies 6
Introduction 7
Bachelor Cook 9
Bear in Camp 11
Buck Camp at Rattlesnake Lake 13
Buggy Ride 17
Can't Build a Fire 21
Chinook Salmon 23
Civet Cat Trap 25
Clancy 27
DuBose the Saddle Shooter Around 1925 29
Elmer Holmes and Cliff's Buck 33
Escape From Alcatraz 35
Ethel Lake Trip 39
Facey's Rock Pile 41
Granddad and Aunt Nellie Bryan Burton 45
Help Yourself to the Mustard 47
Home Brew 49
Horse Wreck 51
Isaac and the Bear 55
James Bryan 57
Jones Family Christmas Letter 61
Klamath National Forest Fires, 1987 65
Mad Cow 67
Milk Parlor at the Bryan Ranch 71
Moonshine 73
Old Lady Wey 75
Old Timer Telling a Hunting Story per Elmer Holmes 77
Pumpkin Jar 79
Rope a Pig 1944 or '45? 81
Second Valley Bucks 83
Sheriff's Posse Ride 87
Shot in the Air 89

Section 2 Poems

- Abe 93
- Always Close the Gate 95
- Bink and the Mexican Piss Patrol. 97
- Black Bear. 99
- Butch 101
- Butch's New Heart 105
- Cowboy Poets 107
- Denny. 109
- Gus and Call 113
- How Come You Carry a Gun? 115
- L. A. Flats 117
- Locking the Gate 119
- Lucy 123
- Monte. 125
- Ole Smoky 127
- Physical. 129
- Rodeo Dance. 131
- Rolling a Smoke in the Rain 133
- Roundup in the Snow. 135
- Skipper 139
- Tax Poem 141
- Valentine's Day Gift. 143

Mike Bryan Biography

Mike Bryan lives in Northern California and is a fourth generation cattle rancher and a licensed mule packer and guide. He has been writing cowboy poetry and performing with his fellow poets and singers, The Siskiyou Seven, for over 25 years. His family ranch was established prior to the Civil War and provides a rich source for many of his stories and poems.

Skip Hanna Biography

Harry "Skip" Hanna grew up in Scott Valley, CA on his family's ranch and showed a talent for drawing at a young age. He graduated from Etna High School and Oregon State University, then served in the U.S. Navy. After the Navy he returned to Scott Valley, where he owns and operates a cattle and hay ranch. He has been a professional cartoonist for many years and has contributed his talents to numerous local organizations.

He is married to Dawn, has four children, eleven grandchildren, and two great-grandchildren.

Introduction

When I first started writing down these stories, I was mainly concerned they not be lost for the generations to follow me. They were stories that my grandparents and parents and friends and neighbors had told me, and since I thought they were pretty funny or maybe had some interesting history, I decided to try to preserve them. Then I thought I ought to add in my poems, and take the whole thing to a printer and give a few copies to my kids and grandkids. But one day I was visiting my next-door neighbor, Skip Hanna, after he had hip surgery and I told him about my project. He simply said: "Would you like me to illustrate it?"

Since Skip is a professional cartoonist, and one of the best in my opinion, I jumped at the chance, and all of a sudden what started out as just a small project for my immediate family, turned into this book. I hope you enjoy it as much as I have writing it and working with Skip. His cartoons have taken it to a whole other level.

Mike

Bachelor Cook

Some ladies were talking with a bachelor and the subject of cooking came up. He said he wasn't much of a cook. One of the ladies said: "Why don't you get a cook book and just follow the directions?" He said he had a book and tried to follow the directions, but never had any of the ingredients.

"Here, I'll show you," he said. He picked up a cook book, opened it, and started to read the instructions for a cake. "The first thing it said was: 'Take a clean bowl.' See, I'm stopped right there."

I was staying the night with Bill Roberts, a bachelor, in his cabin down river. I got up early in the morning and started a fire in the wood stove and put coffee on to boil. Bill was outside feeding horses and mules, and when he heard me stirring around and smelled coffee, he hollered in to start the bacon cooking, he'd be right in.

I found the cast iron fry pan and when I looked at it, it had not been washed out and there was bacon grease in it. The white grease had mouse tracks in it. I hollered to Bill: "There's mouse tracks in this fry pan!" Bill said: " Don't worry, they'll go away as soon as you put it on the stove."

Bear in Camp

We were camped on a little lake in the Trinity Alps. The timber came to the edge of the lake on the west side. It was full of deadfalls so horses couldn't get out on that side. There was a good meadow above the lake, no way out. The only way out was through camp. We put up ropes and saddle blankets for a barrier.

Frank was the wrangler, he tied her all up neat. He says: "They'll be all right behind those ropes, we can sleep easy tonight, and won't have to graze stock in the morning." The lead gelding had a bell (no mares) so if they did take off we could hear them.

We were all asleep when the bell started to clang, horses and mules were pounding around the small meadow. Frank was up and running around the outlet of the lake toward the timber on the west side, as he heard timber crashing there. He was bare foot and naked but he made a pretty good run across the meadow. He hit a pothole in the meadow and went clear to his neck in the cold muck, but climbed out and kept going. The crashing faded away ahead of him and he came back to camp cold, muddy, and wet.

"I couldn't catch whoever it was, we might as well wait 'til daylight so I can track whoever got away. I don't see how they could get through all those deadfalls." At daylight they checked the remuda and all were there. Frank retraced his midnight run and found where a bear had crashed through the brush and logs. Apparently he startled the stock and they ran and the bell scared him, so he ran off through the trees. Then, of course, he had a naked man chasing him.

They asked Frank what he would have done if he had caught the bear. He said: "Hell, I would have rode the s.o.b. back to camp, my feet were sore."

Buck Camp at Rattlesnake Lake

All guided trips to the mountains have the potential for problems because of the terrain, working with animals, weather, and other variables, but there is one trip in 2001 that stands out for its problems and frustrations.

On September 26 I had a hunting party booked for a trip to Rattlesnake Lake in the Trinity Alps Wilderness Area. I only used the trailhead for Rattlesnake Lake during deer season. I had used it in 2000, but was doped up because of a horse wreck the month before and did not remember how to get to the trailhead.

After wasting time wandering around, my assistant packer, Kathy Wood, and I finally found the right road. When we got close to the trailhead the Forest Service had a berm pushed up on the access road. Their property survey had found that part of the old lot was inside the wilderness boundary, so they more or less abandoned the project. The spot they were calling the trailhead had room to turn a truck and trailer around, but only if there were no other rigs in it. There was one rig parked, so we could not turn around. We walked in to check this out before driving in and we managed to turn around in the road and park alongside the road below the parking lot.

We had two truck and trailers and eleven head of horses and mules. We had to cut swamp alders out of the way to tie up for overnight. Stock water was at a culvert 100 yards down the road with very poor access. We made camp and fed hay to our stock and made ready for our morning trip.

Our buck hunting clients made camp in the parking lot as they could turn pickups around there. The next morning we fed and watered our stock, saddled, and loaded the clients' gear on the mules. Out of the five hunters, three were riding, two were walking, and we had six mules to pack. We ended up with eight mules packed, one rider, and four walking as they had brought so much gear. Normally we try to keep our loads to 150 pounds per mule, but this trip we had 200 pounds per mule.

We took the pack string and the clients into Rattlesnake Lake. Rattlesnake Lake is very small. In a good year, it is little more than a mud hole. That year it was all but dried up. There is a spring at the campsite that normally runs into a pool that supplies enough water for stock to drink. That year the spring barely ran enough for camp water, and the pool was dry. We tried to water the stock at the lake, but it was so muddy we could not get them to drink. The only water they had was at the trailhead before we started to pack up, and again at the trailhead in the late afternoon as we went in and out the same day.

On Thursday, October 4, Kathy and I drove back to Harry Hull trailhead. Lee Wood, Kathy's husband, drove a third truck and trailer because we needed fifteen head of stock to bring the party out. There were no rigs in the parking lot so we were able to turn around in the lot. Lee stayed at the trailhead and Kathy and I went into Rattlesnake Lake that same day so we could get an early start the next morning. There is not a good place to feed stock at Rattlesnake Camp, so we packed in grain and hay.

When we got to camp there was no stock water at the spring, so we unsaddled everything and led a few to the lake, with the rest of the stock following. The mud was brisket deep to a tall mule before they could get to very muddy water. None of the animals would drink very much, some not at all.

As I said, we had lots of hay. When I had gone by the barn the day before I had just thrown on three bales that were handy. The hay was third cutting grass/alfalfa, but heavy on the alfalfa so it was pretty hot feed.

The gelding I had been riding all summer, Twist, had lost so much weight I had decided to give him a rest. I was riding my wife's gelding, Tio, who was about seventeen years old and as well broke as any animal we ever owned. He was a quarter horse and had been a good horse. He was a little too much for most dudes, I mean, clients, but great for anyone who knew a little about horses.

We tied up the stock and fed some hay and grain that night. In the morning they would not go into the mud hole to drink but we fed more hay and grain. The hunting party had killed five bucks and a bear. The bear head and hide did not smell too sweet, as they had killed it the first day they hunted. We put them in an ice chest, but you could still smell them.

By the time the clients broke camp and got their gear together and Kathy and I got it on the mules, it was almost noon. We had our fifteen head of stock, two for us to ride, three clients mounted, and ten head of pack stock. In 2000 when we brought this same party out, there were four packers. This year I had another man lined up, but he had to cancel out at the last minute, so there were only the two of us.

Kathy stepped on her riding horse, Crooked Nose, and had five mules strung behind her. I was going to ride in the lead with my string, then Kathy with hers, then the three mounted clients. When I stepped on Tio, he went sideways clean across the campsite, broadsided and swiped Kathy and Crooked Nose, and had all the mules balled up. Kathy said: "I'm out of here before we get in a horse wreck," and she took the lead.

The trail is narrow and steep when it first leaves the campsite. I had all I could do to keep Tio from crowding Kathy's last mule off the trail.

My five were crowding into me and the whole bunch was acting crazy. Past the steep part of the trail there was a lot of brush growing over the trail, in some places head high to a man on horseback, other places just knee high, but thick.

The mules kept boogering and acting up. They broke breakaways repeatedly. (A breakaway is a light rope or twine tied between the lead rope and the next mule's saddle. It is intended to "break away" if a mule falls or goes over a bank so the whole pack string is not lost.) At one point when I got off to retie a breakaway, I got bumped by a mule and ended up head down with my spurs hung up in the brush. One of my clients said: "Mike, you okay?" as all he could see was my boots and spurs sticking out of the brush. I managed to get out of the brush, get back on Tio, and go another ¼ mile or so, when we got into yellow jackets, which blew up the pack string again. We kept on having mini-wrecks all the way out.

I was bound and determined not to lose my temper with Tio and my mules, as I had had a bad horse wreck the year before mainly caused by my getting mad. Kathy's horse and her string were acting up, also. Even though Tio was an exceptionally well-broke horse, something sure had him juiced up on this trip. It was probably a combination of bad water, bear and buck smells, yellow jackets, and too much hot food. Whatever it was, after about two hours of restraining myself, I was so mad I figured if I stayed on him one of us was going to end up dead, so I got off and led him most of the rest of the way out. He and the mules still acted up, but seemed to be under a little more control.

At about the time I started to walk, I went into the lead and Kathy followed me. That seemed to help calm Tio down a little, although at some point when I was walking he got into the brush and hung my saddle gun up and, unbeknownst to me, broke the strap tying the scabbard to the saddle, dumping the rifle in the trail. Fortunately, one of the clients saw it and picked it up.

I finally got back on for the last mile or so. We unpacked the stock, and then had to lead them one at a time to water. We then loaded the trailers and headed home. About three miles down the road, I looked in the mirror and saw a mule's hind leg sticking out the side of the trailer through the small opening about five feet up from the floor. I stopped, unloaded the back mule, then the mule with his leg out pulled it back in. I have an aluminum trailer and as he pulled his leg in, he tore the entire hide and flesh off the cannon bone from the hock to the fetlock. Three months later, it was still an open wound, but eventually it did heal.

If you are waiting for a happy ending, I guess the fact that nobody was killed or hurt, the clients had a successful hunt, and the mule healed up is happy.

Buggy Ride

Tom Edwards and I were 15 or 16 years old. We each had a driver's license when we were 14. Tom had a Model A Ford car. My cousin, Jim Holmes, Tom, and I were all in Tom's Model A. We had gone squirrel shooting up Rance (now Shell) Gulch on a Saturday morning in the spring.

There was an old cabin and outbuilding there, the Rance cabin. They were abandoned as Ervin Rance had died several years before. There was an old buggy in the shed and we decided it would be great sport to ride the buggy down a hill. The buggy did not have shafts or a tongue, so we tied a rope to the front axle and towed it as far up a hill as we could go with the Model A.

The rope was tied to the front axle, one end on the right side, one on the left side. The middle of the rope was looped over the single seat and into the small box behind the seat. Two people could sit side by side in the seat. The third person, Tom, sat in the bed behind the seat and was the one to steer the buggy.

The only problem was he could not see ahead because the two guys in front blocked his view. So, in order to steer, the guys in front would say "right" or "left" to guide the driver. There was a fence to cross which had a hole about 30 feet wide we planned to drive through.

Jim and I were in front. The buggy had a brake, Jim was the brakeman. Of course, if you used the brake, the ride was not going to be very fast. We were not about to use it unless it was really necessary.

We started down, slow at first. "Right," Tom pulled right, we went right, no problem. "Left," Tom pulled the left rope, we went left. We were picking up speed, "right," Tom pulled the right rope. The buggy went up on the left two wheels as he turned a little too far to the right. By this time we were zig-zagging back and forth, too far right to go through the hole in the fence, then too far left. The buggy was going up on two wheels with each turn and we were going faster and faster.

At some point I decided that I would be better off out of the buggy than in it, so I jumped out, hit the ground, went down and rolled, and somehow the buggy ran over me. This has always been a mystery to me as to how that happened, but it did. When I went out, Jim braked to a stop, and the ride was over.

My only injury was a sprained wrist. This would not have been too bad except that I was supposed to drive tractor that afternoon, pulling a spring tooth harrow on one of the alfalfa fields. The tractor had tricycle wheels and no power steering, so both hands were needed to hold on to the steering wheel.

I didn't tell my parents about my wreck as I figured that might curtail further adventures. So, I drove the tractor all afternoon in considerable pain, especially each time those front wheels hit something that jerked the steering wheel around.

If there is a moral to this story, I think I missed it because I am in my 70's and I still have a wreck every year or so.

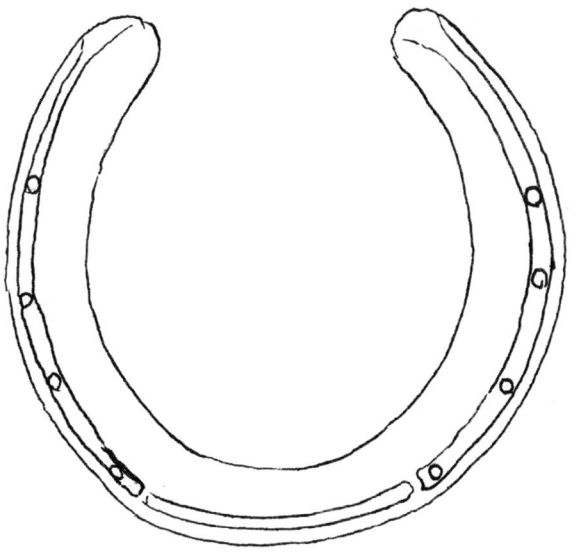

Can't Build a Fire

When I was a child my parents and I lived in an old house on the family ranch. The house was heated by a fireplace in the living room and a wood-burning cook stove in the kitchen. My dad would build a fire in the kitchen stove every morning. I was always up early to watch the procedure.

Dad used old newspapers as a fire starter, then kindling, then larger pieces of wood. When Dad picked up the newspapers for fire starting, he would look at them, usually reading something he had missed or maybe he had never read that paper at all. To my three-year-old brain this was an important part of the fire starting procedure.

One morning my dad was later in getting up and I was up and running around. I went into my parents' room and my dad said: "Why don't you build the fire?" I said: "But Dad, I can't read."

Chinook Salmon

When I was in first or second grade, I rode the school bus to and from the grammar school in Etna. The bus driver was Sam Potter. He was a good driver and he usually drove pretty fast. We called our school bus "the cracker box" because it was not a full size bus. It probably carried 20 or so passengers.

The bus route varied from year to year, but this year we came east on Eller Lane on the afternoon run. Part of the road across Eller Lane parallels Patterson Creek for a mile or so. We could see the creek bed from the bus. One day on the way home from school, the creek was full of spawning Chinook salmon. Sam stopped the bus, opened the door, and said: "O.K. kids, everyone out and get a fish."

The creek bed was pretty flat and only four or five inches deep. We all caught fish with our bare hands and packed them into the bus. When I got home, I had to walk the quarter mile long lane to my house. I had the fish on my back, holding the head by the gills over my shoulder. The tail dragged in the dirt. I sure was proud of that fish.

By the time salmon get that far up the creeks they are pretty beat up. I'm sure this one was no exception. The meat was mostly brown and the skin was missing in patches. But my mother cooked some for dinner and I thought it was real good.

How many of today's "rules" were broken by us kids and the bus driver? I cannot imagine a bus driver today stopping and letting kids run up a creek chasing salmon, yet neither we nor our parents gave it a second thought. Of course, I suppose they would have criticized Sam for going into the roadside ditch with the bus to try and run over a coyote, which Sam did, but that's another story.

Civet Cat Trap

When I was about ten years old we had lots of civet cats on the ranch. A civet cat is a spotted skunk. They like to make their homes under houses or granaries or in the lofts of farm buildings and are notorious egg suckers and chicken killers. When disturbed, they spray a strong odor just like a skunk. The smell sticks to everything, including your clothes.

One day, I decided to set a civet cat trap in the "cat hole" in a granary loft door. The door was off a 6'x8' platform about eight feet above the ground, so I could see it pretty well at a distance. The next morning I looked at my trap and I had a civet cat in it. I wasn't about to go up to that trap and get sprayed, so I figured I would shoot the civet cat. If he were dead, he couldn't spray me.

So, I got my .22 and took careful aim, from some distance back, and shot the cat. Well, the lower half of the civet cat quit moving but the upper half continued to move, not just a little bit but pretty active. This was fairly unnerving to my ten-year-old brain. I couldn't figure out why the bottom half was not moving, but the top half was.

My dad was not home so I got my granddad to help me out. I told him I had shot a civet cat in a trap and he was half dead. My granddad said: "Shoot him again." I said it was really strange but the bottom half was dead and the top half was alive.

Granddad came to see this very strange civet cat. Neither one of us wanted to get very close, so from a distance he couldn't figure it out, either. He said: "Well, go ahead and shoot the top half." I did, it quit moving, and after waiting awhile, we cautiously approached the trap. There were two dead critters in the trap. The first shot I killed one, the second shot, I killed the other one. Mystery solved.

Clancy

My dad, Frank Bryan, owned a Saddle Bred mare. He had her bred to several different studs during the time he owned her. One of these sires was an Arab. The mare foaled a blaze-faced, white socks, copper-colored colt. Dad named him "Clancy." He certainly was flashy. As with most of the foals we raised in those days, he never was well broke, we basically just rode them.

Clancy was about 16 hands and very athletic. We rode in the mountains a lot; so he was used to steep, rough country. My dad was riding Clancy on a buck hunt we went on one fall in the Marble Mountains. We had stopped on top of a ridge to "parley." Clancy, as usual, was faunching around when he stepped on and then fell off a four-foot high rock and landed on all four feet on top of a mountain mahogany log about 8" in diameter, and a crooked 8' long.

He balanced there, tipping this way and that as the log rolled back and forth under him

Then he jumped straight up in the air and back on top of the rock. (It was too "bluffie" to go downhill.) My dad never left the saddle; he couldn't, no place to go.

Same trip, different day, Dad was riding Clancy, leading another horse named Casey. Casey was an older horse and not real quick. He had side bones and didn't move very fast. We got hit by yellow jackets. Clancy jumped and bucked to the downhill side and turned about 90 degrees to the trail straight downhill. My dad had tied a loop in the lead rope and had it over the saddle horn. The "jump, buck, turn downhill" jerked Casey down on his side.

So, (I know you see this one coming), my dad was pinned in the saddle. No way out, with Casey a dead weight and being tied hard and fast over his leg, under the cantle, and back to Casey. After about three good jumps, which took the wreck 60 feet down the hill, Dad got Clancy turned and got his leg free and baled off.

Dad had some cuts and bruises, Casey had some cuts and scrapes, but Clancy seemed to be none the worse for the wreck. Oh, by the way, Dad never tied a packhorse or mule hard and fast to his saddle horn again.

DuBose the Saddle Shooter Around 1925

In the 1920's the Mathews clan ran cattle in the Marble Mountains. They rounded them up in the fall and brought them down to Scott Valley to winter them. The fall roundup coincided with buck season, so they usually hunted and rounded up cattle at the same time. As with most of the ranches at that time, it was a big deal. Neighbors and friends were invited to participate and family members came home to hunt and help.

Dick Mathews' oldest son, Jim, was at West Point and he came home with two other cadets, DuBose and another man whose name I can't remember. My dad, Frank Bryan, was a good friend of Jim's and was along on the trip.

They had a large camp with a lot of horses and mules. My dad, Jim, and DuBose decided to "spike out" overnight some distance from the base camp to have a better chance at the hunting, away from all the noise and confusion of the base camp. Since this was a spike camp, they did not take their heavy bed rolls or a lot of gear. They had their saddle horses, a small amount of food, and a frying pan. They all used wool army blankets for saddle blankets, and planned on putting one of these on the ground, with two over the top, and all three men sleeping in one bed.

They ate their supper, had a good fire, and were thinking about the cold night and the three blankets. Jim was thinking it would be warmer to be in the middle. Frank was thinking the same thing. They sat around a little longer and DuBose said: "Guess I'll sleep in the middle." Frank and Jim didn't say anything for a while. After all, he was a guest and they didn't want to treat him bad.

After a short time, Frank recalled "that time in Salmon River when three miners were in one bed and a bear came into camp and mauled the man in the middle." Jim remembered "the time near Happy Camp when three Indians were in a bed and a mountain lion dragged off the one in the middle." They each recalled several incidents of three in a bed and the one in the middle being killed or mauled or dragged off.

After all these stories, DuBose said: "One of you guys can sleep in the middle." Of course, they said: "Oh no, you're the guest, it will be warmer there." By this time, DuBose was pretty worried. He asked: "Nothing would really come into camp, would it?" Jim and Frank assured him that would be very unlikely, besides, if something did come in ol' Brownie, the dog, would get on the bed and "he would not bark, just growl real low in his throat." This was a total fabrication, they had no idea what ol' Brownie would do.

DuBose was not totally convinced that he was safe. They finally bedded down, but DuBose insisted he have his rifle with him in bed. Well, as luck would have it (good or bad), ol' Brownie jumped on the bed in the middle of the night and growled real low in his throat. DuBose sat up fast, gun in hand, the fire had died way down and was just a glow. But it was enough light to reflect off two of the silver conchos on DuBose's saddle across the fire on a log. He levered a shell into his rifle and was about to shoot his saddle "between the eyes." Jim and Frank got his rifle away from him and showed him what he was about to shoot.

Sounds like "end of the story?" Not quite. When they got back to the base camp the next day Frank and Jim, of course, told the story. They embellished it a little to say DuBose <u>had</u> shot three holes in his saddle before they got his rifle away from him. Everyone else in camp went along with the story. DuBose got real mad, grabbed up his saddle and showed everybody that it had no bullet holes in it.

Dick said: "That's too bad, ruined a good saddle." Everyone else pretended the saddle was ruined. DuBose was almost convinced that he <u>had</u> shot three holes in his saddle!

When Jim and DuBose and the third cadet returned to West Point, the story was repeated and DuBose was forever known as "DuBose, the saddle shooter."

Sadly, all three of those West Point men were killed at Normandy in June of 1944.

Elmer Holmes and Cliff's Buck

In the 1940's, my dad, Frank Bryan, and his first cousins, Elmer and Cliff Holmes often hunted bucks together. One of the Holmes brothers' nephews came up from the Bay Area and wanted to go hunting with them.

Frank, Cliff, Elmer, and the nephew, Jack, went hunting on Schoolhouse Hill*. They had hunted this area for years and knew every canyon and ridge. They had agreed to meet at a certain point at 3 p.m. before going downhill to their pickup parked at the bottom of the hill. All showed up at the given time except Cliff. They waited a while for him, then Elmer said: "I saw Cliff below me a little while back. I thought I heard a shot; maybe he killed a buck and needs help. I will go check and be back shortly."

Elmer went back around the hill and found his brother just finishing dressing out a buck. Elmer said: "You need any help getting him out?" Cliff said: "No I can pack him okay. You go and tell the others and I'll meet you back at the pickup."

Elmer went back to where the rest of the party was but told them he couldn't find Cliff, but that he probably was headed to the pickup so they might as well wait for him there.

As they were going back, Elmer cut Cliff's track. He pretended he didn't know it was Cliff's and started to "read" the sign, mainly for Jack's benefit. "Fresh man track, about a size 8 boot, 5'11" tall, (see the length of the stride), probably about 175#. Man was carrying a buck on his back, looks to be about 120# or so. See how his boot heel cut deep when he crossed that gully. Three point on the left, four point on the right, track's about 20 minutes old."

When they got back to the truck, Cliff was there, 5'11", size 8 boots, 175#, with a 120# buck, 3 on the left, 4 on the right. Frank knew that Elmer was pulling Jack's leg, but Jack was real impressed that Elmer could read sign so well that he could tell all he had by just following a man's footprints.

*There are several hills in Scott Valley that are referred to as "Schoolhouse Hill." This one is east of the old Franklin schoolhouse where Tom Hayden now lives.

Escape From Alcatraz

On June 11, 1962, brothers Clarence and John Anglin and another prisoner, Frank Morris attempted to escape from the federal prison, Alcatraz, or "The Rock." Alcatraz is located on an island in the middle of San Francisco Bay and was reserved for the most hardened criminals, or the ones who had previously attempted to escape from other prisons. It was considered next to impossible to escape from Alcatraz because of the distance from the shore, the freezing cold temperature of the water surrounding it, and the strong tides that make swimming extremely difficult, especially for men who have been prisoners. All three of these men had long criminal records and the Anglin brothers had a history of attempted prison escapes.

I was a U.S. Coast Guard diver stationed at Captain of the Port (COTP) on Fisherman's Wharf in San Francisco at the time of this incident. I was an officer with enlisted men on my diving crew. A massive search was launched, coordinated by the FBI, which used the COTP office as their base because we had radio contact with all the Coast Guard ships, boats, and aircraft. We also had a radioman monitoring all the radio traffic on "2182" the national distress frequency, which was used by all fishing and small boats twenty-four hours a day.

Besides the Coast Guard 40 ft. patrol boats, there were larger Coast Guard boats on alert outside the bay. The local police boats were all searching along with Coast Guard helicopters and planes. The FBI men would ride in the Coast Guard helicopters during the search. The local fishing fleet was also on the look-out for the prisoners during its daily trips outside the Golden Gate.

The search boats that were circling Alcatraz looking for some sign of the escaped prisoners picked up some pieces of prison denim. They also found some crude life jackets that were made from prison rain gear, as well as remnants of the raft, paddles, and a bag containing the Anglins personal effects.

The Coast Guard divers were assigned the task of searching closely around the island and looking into the caves for signs of the escapees. We had small 40 ft. patrol boats manned by the Coast Guard boat coxswains who were used to going in close to shore. The COTP captain was Captain B.P. Clark. He was more concerned for the safety of his men than in finding the prisoners, and did not want his divers to be in danger.

My plan was to have a small inflatable lifeboat on the 40 ft. patrol boat. When we needed to get closer than the 40 footer could go, Ed Hanson, who was left-handed, and I, who am right-handed, would put the small lifeboat under one arm and hold a 1911 .45 Colt pistol in the other dominant hand as we swam into those caves.

When I presented my plan to Captain Clark, he said that was NOT a good idea and he did not approve, so we did not pack the .45's when we searched the caves. As it turned out, we did not have to swim into any of the caves as we could see in from the 40 ft. patrol boat with powerful flashlights. This search did not turn up any sign of the escapees.

When the Anglins and Frank Morris were not found by the initial search, the USCG divers were assigned the task of searching the water around Alcatraz. As I said, the Bay currents are very strong and at times when the tide is in full flow it can run at over three or four knots. We had to time our diving to slack tide times only, which only gave us maybe two hours each day. We could not search at night.

The prison guards were watching us as we dove, and when we had to stop because of the strong current, they invited us into Alcatraz for lunch. As we went into the dining hall, the iron prison doors clanged loudly behind us, which made my divers and me quite uncomfortable. But just being inside Alcatraz gave us a weird feeling.

The guards had their own dining area, but were served their meals by the prisoners. All these guys walking around in prison blues made me and my divers a little nervous. The guards were getting a kick out of us military men being so nervous, so one of them mentioned that the cook was a poisoner. That helped a lot to settle us down.

Other than finding some prison denim, we found no other sign of the escapees. The prison denim was not necessarily a sign of the escaped prisoners as inmates often tried to plug up the plumbing by flushing denim down the toilets. It was reported that a Norwegian freighter spotted a body floating in the ocean fifteen miles from the Golden Gate bridge about one month after the escape. He had on prison clothes; a navy pea coat and a light pair of trousers similar to what Alcatraz prisoners wore. There were no other reported missing people during that time period.

It is my personal opinion that the three men did not survive their attempted escape despite many rumors to the contrary. These men were all hardened, career criminals and it is highly unlikely that had they lived they would not have resumed their lives of crime and shown up in the criminal justice system sometime later. No one ever heard anything about them again.

Ethel Lake Trip

I had a client who insisted he wanted to go a long way out so he and his party could get away from all the people. He chose Ethel Lake in the Salmon Mountains. Ethel Lake is twenty miles or so from the trailhead at Mule Bridge. I tried to talk him into something closer, but he insisted part of the reason I did not want to take him that far was that I didn't feel like charging him for two days each way, so we ended up charging him for 1 ½ days in and 1 ½ days out. My helper, John Lyons, and I stayed the night when we took the party in, and went in a day early to bring them out.

They were real nice people, but they were dudes so we ended up taking breaks, stopping for lunch, etc. The trip went well, but by the time we got to the top of the ridge above the lake it was late. The trail is easy and not at all steep until you get to the ridge top about a mile from Ethel Lake. I was in the lead, and knowing they were all novice riders, I turned around in the saddle and said "Lean back and put your feet forward, it gets steep here."

One of the ladies on the trip was small so I had her on a small half-Arab gelding. His name was Whiskey and we mostly used him for a kid's horse. The trail was steep, but not rocky or dangerous except just before it got to the flat near the lake. At this point there was a big Shasta fir with a large root across the trail. The root held back the soil in the trail bed on the uphill side, but just over the root the trail dropped about two feet straight down.

Whiskey stepped over that root and just dropped his front feet the two feet down. This lady had short legs and at this point, Whiskey's back was almost perpendicular. She went off right over his head and rolled onto the flat. I jumped off my horse and went to her. Well, she was shook up and covered with dirt, but didn't seem to be hurt much. I asked her if she was hurt. She said: "I guess I'm okay, but I knew we were in trouble when you said: 'Lean back and put your feet forward.'"

It turned out she was okay, but when we came back to pack them out, the whole party hiked to the top of the ridge and John and I led their horses up. It was a long trip out to the trailhead but we didn't have any further mishaps. Strange thing, though, those folks never booked with me again.

Facey's Rock Pile

When I was in high school my friends and I used to like to hike and run around in the hills close to home. I was about 16 or 17 years old, the year was 1952 or '53. One spring Saturday we decided to explore the caves on "Facey's rock pile." Facey's rock pile is the limestone hills east of East Callahan Road about four miles north of Callahan. Access was across private land owned by the Faceys. This is a limestone formation that rises about 2000 feet above the valley floor.

There are several caves there and my dad, Frank Bryan, told me about one that had a funnel shaped opening and went straight down for a long ways. He said when he was a kid they would roll rocks into the funnel and hear them rumble for several minutes. He told me how to get to it.

There were five of us in the group: Tom Edwards, John Akins, Lorenzo Davis, Paul Black, and me. We found the cave entrance, which was just as my dad told me, a funnel shaped top narrowing to a small hole and then opening into a larger cavern. We rolled some rocks into the funnel, but they just fell a short distance and did not "rumble for several minutes." They just stopped.

We cut a juniper pole, put it across the top of the funnel, tied a rope to the pole and dropped the end into the cavern. I was elected to go down the rope. I had a dim flashlight in my pocket. When I got through the narrow opening at the bottom of the funnel, the cave opened up, so I had the rope but no wall I could reach. I was hanging in mid-air on the rope with no wall or place for my feet. I went down the rope hand over hand, and when I got to the knot at the end of the rope my feet would not touch the floor of the cave. I could not let go with one hand to get the flashlight out of my pocket, so I was hanging on the end of the rope in the dark, wondering how far it was to drop, with the possibility of rattlesnakes being curled up, waiting for me, flashing through my mind.

We had looked down the hole with the flashlight before I went down and we knew it was not too far to the floor, but I could not touch it with my feet. I hung on to the rope as long as I could, but knew I would have to drop eventually as my "friends" up top could not or would not pull me up. I finally let go of the rope, prepared to fall I did not know how far. The "fall" was only a few inches, but the floor sloped down steeply and I fell and slid to a rock jumble, but with no damage and no rattlesnakes.

The rope was hauled up and a short piece added to the end so I could reach it to get out. We thought maybe there had been so many rocks rolled in the entrance over the years that they had plugged the way down. We spent some time trying to dig down and unplug the hole. By this time,

Tom Edwards had joined me in the task. There was only room for one of us at the very bottom of the sloping cave floor, so we took turns prying out rocks, handing them back to the other one to stack out of the way.

It was about noon by then and the guys topside finished their lunch. John Akins took all the used lunch wrap waxed paper, rolled it into a ball, lit it on fire, and threw the ball down into the cavern on top of Tom and me. He yelled, "Fire in the hole" and we looked up to see a ball of fire descending on us. I was head down at an angle trying to lift a rock out of the bottom and could not get out of the way. We did not know the "fire ball" was just waxed paper. I looked up to see nothing but fire falling toward us. I tried to get out of the way, but had no place to go. The waxed paper burned up and no harm was done.

We did not get the hole unplugged and we gave up after a short time. We came up out of the hole, collected our rope, and went on to the next cave.

That morning when we had gone to pick up Lorenzo, he wasn't ready to go. We went into his house to wait for him. His father, Bert, was there and Lorenzo asked him if he could take the pistol. Bert said, in his real low, slow voice: "Yeah, but you better take the heavy loads, Sonny, there might be a bar in one of them caves." The rest of the guys were all carrying pistols, but they were .22 caliber. This hog leg Sonny (Lorenzo's family called him Sonny) was packing was an old Colt single action Army in 44-40 caliber and it had an 8" barrel and weighed about four pounds. It hung clear to Lorenzo's knee. The cartridges in their leather loops were green crusted and looked as if they been in that belt for years.

We were all shooting rocks, cones, sticks, etc., with our little .22's. We teased Lorenzo about being able to hit anything with that old hog leg. Tom Edwards was wearing a red hunter's felt hat with a fancy hatband. Tom said: "Here, 'Renzo, I'll throw up my hat. See if you can hit it." He threw his hat in the air and 'Renzo not only hit it, but that big .44 caliber bullet went through the hatband and sweatband on both sides of that hat and ruined it. Tom picked up his hat and put his fingers through the hole. All of us figured Sonny had made a lucky shot and so we tried to get him to shoot again. But he refused, saying, "I don't want to waste any more ammo."

We didn't tease 'Renzo anymore about being able to hit anything with that old hog leg, but "better take the heavy loads, Sonny, there might be a bar in one of those caves," became a part of our vocabulary.

On our way down the hill that afternoon going back to our car, we ran into a bobcat. He was part way up an oak in the crotch of the tree, snarling at us. We were pretty close to him before we saw him and he scared us. Two or three of us emptied our .22's at him. A couple of the bullets actually hit and killed him. It turned out he was in a leg hold trap that had been set by a Federal trapper.

Not being real impressed by whose trap it was, we took him out of the trap and Paul Black put him over his shoulders and took him home to skin him for the pelt. That cat ran urine on Paul's leather coat all the way down the hill. When Paul came to school Monday morning in that coat, it had a pretty strong smell.

Granddad and Aunt Nellie Bryan Burton

My grandfather, Charlie Bryan, was born in Scott Valley on the Bryan Ranch on Eastside Road on September 9, 1867. In the early to mid-1960's he was living in Etna with his daughter, Helen Bryan Ball, so he was well into his 90's at the time of this incident.

Granddad was a great story teller, so Lynne and I thought it would be a good idea if we could record some of his stories. We didn't tell him we were recording because we knew he would feel intimidated, so we set up the machine where he couldn't see it.

Aunt Helen had a party and invited family members, including Granddad's sister, Nellie Bryan Burton. After dinner, we got Granddad stated telling stories. It didn't take much effort as he loved to tell stories. I don't remember what the story was about, but it had a cat in it. Aunt Nellie was sitting in a rocking chair listening to her brother spin this tale. The more he got into the story, the faster she rocked that chair. Then she started tapping her foot, and got a mad look on her face.

When Granddad finished the story, she said very emphatically: "That's not the way it was, Charlie! The cat was black, not white. And it was a she, not a he." And she had several other things about the story she disputed.

Granddad and Aunt Nellie argued so long about the details of the story that he got mad and that was the end of the story telling. We never did get his stories recorded.

Help Yourself to the Mustard

My grandfather, Charlie Bryan, along with a hired man, was riding for cows in the East Side Scott Valley hills. Along about noon he came to an old cabin with a man living there alone. There was no road to this cabin and the man didn't get too much company, if any.

Well, he was happy to have visitors and invited my grandfather and his hired man in for lunch. As I said, he didn't get much company, so he had to hustle some to find three plates and three spoons. He threw a stick of wood in the stove and heated up a pot of beans. When the beans were hot, he set them on the table and said: "Have some beans."

My grandfather spooned beans onto his plate, but the hired man said: "No thanks, I don't care for beans." The host said: "Well then, help yourself to the mustard."

Ever since then whenever someone in our family doesn't like the food that's served, we tell them: "Well then, help yourself to the mustard," and they usually change their minds.

Home Brew

During Prohibition my father and his brother-in-law, John Ball, went hunting with John's brother-in-law, Dell Tabor. My dad made home brew and took some along to drink after the hunt. Dell Tabor was a "Pro-Hi," (teetotaler), so they decided to hide the beer under the back seat of the two-door car they were driving.

The car sat in the hot sun all day while they were hunting. When they got back to the car after the hunt, Dell got in the back seat, which was small with not much room. He said: "Hand me those guns, boys, and I'll put them back here, but make sure they're unloaded." Frank and John unloaded the guns and started to hand them to Dell. Apparently, Dell had sat down hard on the seat and just then the hot home brew blew up.

Dell thought he had been shot by one of the guns. He said: "I told you boys to unload those guns, I've been shot in the leg, I can feel the blood running down my leg!" Frank and John had to "fess up" about the beer. Of course, they had no beer to drink, either.

Horse Wreck

Bryan and Sherman Packing was a mountain guide service started by Bink Sherman and myself in 1988 after we had worked together on the 1987 wildfires here in Siskiyou County. In August of 2000, we had a large party of bow hunters to bring out of Union Lake, up Union Creek, a tributary of Coffee Creek in the Trinity Alps. Bink had another trip that day, so he was not with us. We had twelve or fourteen head of pack mules, plus a bunch of riding stock. I was riding a sorrel thoroughbred I had bought in the spring. He was a little goosey.

My son-in-law, Jim Morris, was riding an Appy gelding that wasn't much better. I rode lead going in and Jim fought the App because he wanted to be in the lead all the time. So Jim rode lead coming out and I was fighting the sorrel who also wanted to be in the lead. He wanted to go so fast that the mules couldn't keep up, so they would break away. I would have to stop and fix the breakaways and we fell behind the lead. My good temper had been somewhat left behind, also.

As in any conflict, it is never all one way or another. In this case, the knot-headed sorrel was not blameless, but neither was I. I lost my temper. I have a little Irish in my genetics, so that may be hard for you to believe. I certainly need to take a large share of the blame for what was about to happen. If you have never been on a hot-blooded horse or never lost your temper, this may not mean much to you. I was a little rough on the horse – using spurs more than needed and raising my voice. I may even have questioned his ancestry.

We had to cross Union Creek on a bridge and the approach was about a two-foot step up. I had to check my horse so each mule had time see the step. This about drove an already hot-blooded thoroughbred crazy. He came off the bridge, spun 180, and went right over backwards. I didn't see it coming. He landed on top of me and the saddle horn, without breaking the skin, went clean to my backbone, rupturing my colon, partially rupturing an artery, and badly bruising my spleen and kidneys, besides knocking all the wind out of me.

My first thought: "I'm a dead man."
My second thought: "I don't want to die."
I prayed over and over: "Lord, give me the strength to get through this."

Dennis Lavey and his step-son, Justin, were with us on the trip. They took the pack stock and the three mounted clients out to the trailhead. Justin was to bring back a gentle mule for me to ride out. At this point I still thought that as soon as I got my wind back, I would be able to ride out. I kept trying to stand and walk, but was unable to do so. Jim, who had EMT training, kept checking me and said he didn't think I had any internal bleeding.

After two hours or so, Jim said: "We need to get you out or make preparations for you to spend the night." I said: "I don't want to spend the night." Jim said: "I'll have to leave you alone then to get help." I said: "Okay, I've got my dog, Isaac."

I don't know how long Jim took getting to the trailhead, but he either galloped or long-trotted the two and a half miles. He then had to drive to Coffee Creek Ranch as the cell phone did not work. He called 911 from there. Mercy Air, Coffee Creek Fire Department Search and Rescue team, and Trinity County Search and Rescue at Weaverville all responded.

As for me, back at the bridge, two hours had passed since Jim left. I was unable to stand or walk without passing out. Somehow I did manage to stagger, in four or five trips, about two or three hundred yards down the trail as I thought there would be no chopper.

I finally couldn't go any more and I lay down in the timber to wait for ground crew rescue. While I was lying there, I both saw and felt a presence near me. It looked like a cloud or fog, but somehow I knew it was the Lord.

My dog, Isaac, was very worried about me, and wouldn't leave my side. At last I thought I heard a chopper and I stood up and <u>ran</u> to a clearing so I could be seen. Remember, I could barely crawl, yet God gave me the strength to run seventy or eighty yards to be seen by the pilot and also to wave my chinks at him.

When the pilot gave me the thumbs-up sign, I collapsed in the trail. The ground rescue crew arrived at that time and they, along with the chopper crew, got me in the helicopter. The pilot had a tough time setting down in that wilderness terrain and was about a quarter mile up the creek from me. I later learned that even though he had night vision gear, he would not have been able to set down in another ten minutes. We lifted off in the dark, with a very anxious dog left on the ground.

When they got me to Mercy Medical Center in Redding, my blood pressure was at 60, I had lost a liter of blood, and I had both gangrene and peritonitis. Part of my colon, in Dr. Bob Ghelfi's words, "was black as coal." The damaged artery had ruptured even more during the flight and I was so close to death that the medics bypassed the emergency room and took me straight to the operating room. Dr. Ghelfi was the surgeon on call and he said later that he cut me open, sorted out the good from the bad, and sewed me up.

I owe my life to the Lord, to all the people involved in the rescue, to my son-in-law, and to the wonderful medical staff at Mercy Medical Center.

Isaac and the Bear

My wilderness guide partner, Bink Sherman, and I had an elk hunting party in the Marble Mountains. We were camped at "Big Rock Camp," a good campsite with horse feed and good drinking water. We were in a meadow just below two small lakes, "Green Granite" and "Gold Granite."

Early one morning, I took the two hunters and their guide on horseback to the top of a canyon. They were going to hunt down the canyon and I would lead their horses around to the bottom. If they killed an elk I would be near so I could take care of the meat. If they didn't make a kill, they would each have a saddle horse to ride back to camp.

On the way up I came to a mule bridge that crossed a small draw and saw a small sow bear lying on her back on the bridge, nursing two good-sized cubs (about 25 lbs.) My dog, Isaac, a border collie/blue heeler cross, trotted across the bridge, jumped over the sow and her cubs, went down the trail fifty feet or so and sat down and looked back. To say this was not typical of a bear/dog encounter would be an understatement. The sow bear, never having had anything like this happen to her before, didn't seem to know what to do.

I stopped my horse and the other three geldings I was leading and just stared at the sow. She never winded me or saw the horses or me. She kept watching Isaac, then finally put one cub down the hill to the right and the other cub eight foot up a tree near me. Now she didn't know which way to turn. She kept watching Isaac, who just sat there, and then she stuck her nose under the cub in the tree. The cub dropped out of the tree and ran across the trail in the direction the first cub had gone. The sow took off after them.

Isaac never moved, just seemed to be casual about the whole thing. My horses were not disturbed. I rode on down the trail, picked up the hunters and guide, and we all rode back to camp.

I, of course, told everybody about my bear encounter and that Isaac was now a bear dog. "He jumped a bear."

James Bryan

My great-grandfather, James Bryan, emigrated from Wexford, Ireland in 1848 when he was twenty years old. He traveled to Green Bay, Wisconsin, joined the U.S. Army and earned the rank of sergeant. In 1852 he was sent to a newly-built Army fort, Fort Jones, in Scott Valley, CA. He came by way of the isthmus of Panama to Fort Columbia in Washington, then down to California. His commanding officer was Ulysses S. Grant, but for some reason Grant never came to Scott Valley.

In 1853 he left the Army and bought one thousand acres of land on the east side of Scott Valley from the government. This was school seminary land, land given by the federal government to the states for the purpose of raising revenue to support public schools. He paid $1.25 an acre. He did not homestead as the Homestead Act was not passed until 1862. For an Irishman from a large family, owning that much land must have seemed like a miracle.

James was a hard worker, unafraid of tackling any kind of job. He owned cattle and ran his ranch, but he also packed beef and other supplies on mules to the miners in Trinity Center and Red Bluff. He owned the Blake Hotel in Etna, and was known as a kind man with a generous spirit. A woman who knew him, Elizabeth Cole, wrote a memory of the early days in Scott Valley and she described him as a short man with brown hair and brown eyes who was "the life of the camp."

A story that I've heard all my life about my great-grandfather shows what kind of friend he was. When he was ranching on his place on the east side of Scott Valley (a place no longer in the Bryan family), his neighbor, Bill Sharp, came to him one day and said: "Guess I won't be your neighbor any more. I was just squatting on my ranch (meaning he had not purchased the land, he was just using it.) Some men found out about it and they are going to buy the land out from under me."

Great-granddad said: "Why don't you beat them to the land office and buy it yourself?"

The law regarding school seminary land stated that it could be purchased for $1.25 an acre, with twenty-five cents an acre down and ten years to pay the remaining dollar per acre. Bill said he didn't have the $250 down payment for one thousand acres. My great-grandfather lent him $250 in gold and a fast horse. Bill rode all night to Sisson, now Mount Shasta, to the land office, beat the other men there, paid the $250, and was James Bryan's neighbor and friend the rest of his life.

This story has been passed down orally through the generations of the Bryans as none of them kept journals nor wrote anything down. There are some variations on the story, depending on who is telling it. One version had the fast horse a fast mule, and another that the land office was in Red Bluff, but the basic story is the same and illustrates the kind of man James Bryan was.

James lived until 1913, and my dad, who was born in December, 1904, remembered him. One thing that made a big impression on Frank about his grandfather was that he was a hairy man with a lot of hair on his back. Frank told a story about the time he and his grandfather were building fence in the Scott River when a carriage with some ladies in it drove by. James and Frank were both naked to keep their clothes from getting wet. James said to my dad: "Go hide in the willows." My father said: "How about you, Granddad?" James answered: "I've got hair covering my body."

Which reminds me of another story about great-grandfather Bryan and how California got the Bear Flag. When his Army infantry company, under the command of Ulysses S. Grant, left Fort Vancouver for Fort Jones, they walked to California. When they got to what is now Grants Pass in Oregon territory, the camped for the night. A bear cub came into the camp looking for a handout. Some of the men caught the cub, and thinking he would make a great mascot, they kept him. The cub grew and became fairly gentle. However, they kept him chained at night so he would not run away or get into their food supplies.

As I said before, James had a lot of body hair. The private who was supposed to chain the bear cub for the night let the chore go until it was almost dark. In the poor light, he caught and chained my great-grandfather instead of the bear cub. The cub, of course, escaped. It is said he pestered the "Fort" for years. James was unchained in the morning. When the story was spread around the rest of California, the idea of using a bear on the California flag was born.

Jones Family Christmas Letter

Well, here we go again, another year has passed in the Jones family. It has been 365 days since our last Christmas letter, and it has been a year full of excitement and adventure.

Our oldest son, Ralph, was tried for the mutilation murder of a young woman and her infant daughter. It was a long, drawn-out trial, but he was ultimately convicted of first degree, premeditated murder and sentenced to be executed. Of course, being in California, he probably will be on Death Row for ten or so years, then be pardoned.

Connie, our oldest daughter, will be 29 this year, doesn't seem possible. She has been working very hard to get her GED after her fourth husband abandoned her and all the kids. The last she heard he was on an oil tanker that sails between the Red Sea and Galveston. He is in charge of the bilge pumps.

Our middle daughter, Lolita, flunked out of Beauty College and ran off with one of the instructors. We do not know where they are, although somebody thought they saw them working as barmaids in a saloon in Yellow Knife, Northwest Territories.

Our youngest son, Ronnie, enlisted in the Navy, but was forced to drop out of Basic Training midway through because his AIDS came back on him and he was unable to keep up with the other men in their rigorous physical training schedule. If his AIDS goes into remission again, he will be able to continue his training in the next class of the San Diego Training Center.

As for "Mom" and "Me," we have been traveling a lot after my accident. I can walk now with the aid of a cane and I have some use of my left arm. Mom is still in a wheelchair but can get in out and out of the pickup we bought to travel in. It's a 1969 Dodge ¾ ton pickup with four-wheel drive, oversized tires, and new paint. We've got a big 8 foot camper on the back and really feel luxurious cruising down the highway at 45 mph. Everybody on the road is really friendly; they all wave and blow their horns when they pass us.

Last spring we went to the Lizard Races in Doyle, CA. What a glorious three days that was. I bet $2 on a big blue-belly in the third race on the second day. He lost by a tail, but we sure had fun cheering for him.

In the fall we went to the Klamath River Blackberry Festival, what a blast. Mom almost o.d'ed on blackberry pie, but there was an EMT there and he gave her the Heimlich maneuver so she did not have to go

to the E.R.
   Well, so long for another year. Merry Christmas and a sober New Year.
   The Jones'

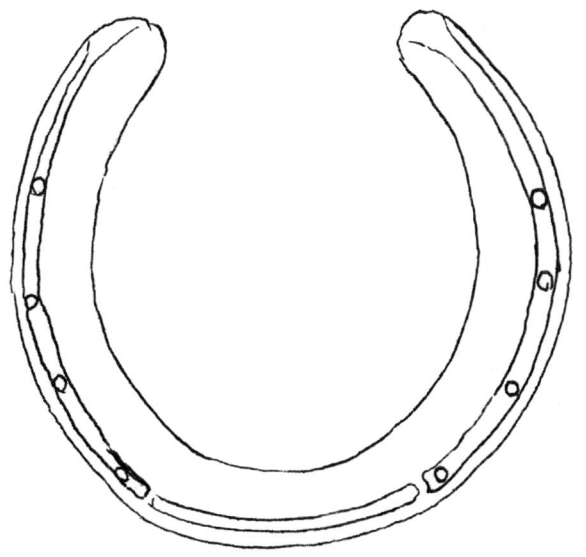

Klamath National Forest Fires, 1987

On September 1, 1987, Siskiyou and Trinity counties received over 3000 lightning strikes. The forest was dry and conditions were right for forest fires. There were so many fires started that the fire fighters could not get to all of them. They burned and burned for days, weeks, and months. Much of the terrain was steep mountainous country, accessible only by foot or horse or mule.

Everybody was fighting fire some place. Bill Coe, Grant Hanna, and I put a pack string together and packed on the fires for over a month. We went where the Forest Service sent us and packed supplies to the fire fighters.

There were fire crews from all over the United States. Many of those crews were "Hot Shot" crews who go all over the U.S., wherever they are needed. A lot of the crews are from Indian tribes, with each crew being from one tribe such as Crow, San Carlos Apache, Blackfoot, etc.

As we packed to the various fire camps and met the Hot Shot crews, I would ask them what tribe they were from. They were proud of their heritage and would tell us their tribe and where their home was. As we rode into one of the fire camps, a young Indian woman came out to meet us and take pictures of the pack string. I asked her what tribe she was from.

Her reply set me back; she said: "I'm Japanese." I took a pretty good razzing about that one. All the guys had to do to get a good laugh from the rest of the crew was to say: "What tribe are you from? Ha! Ha!"

Mad Cow

On January 6, 2004 I was feeding cattle on the east side of the river. There was about a foot of snow on the ground. I had finished feeding when I saw a cow with a new calf. I roped the calf to ear tag him. Cows are very protective of their newborns and this cow was showing signs of being a little snuffy, so I got on the flatbed pickup and pulled the calf up on the bed so the cow couldn't get at me, though she kept trying.

I tagged the calf and while I was checking to see if it was a bull or a heifer it scrambled off the truck and fell to the ground, but still had the rope around its neck. I reached down to take the rope off and right at this point things got a little hectic so I don't remember exactly what happened next. Either the cow hit me and knocked me off the truck or I slipped and fell off; in any case, I was off the truck and on the ground and this 1400-pound Angus cow decided to eat me for lunch.

The first time she hit me, apparently she slipped and dropped her full weight with her brisket onto my rib cage. I was unable to get up and in excruciating pain. She wasn't satisfied with this and kept after me as I tried to roll and scramble away from her. She ended up butting, pushing, rolling, and mauling me a hundred feet away from the truck before she left me for dead.

During the mauling I was in terrible pain and was unable to get up or get away; I thought she was probably going to kill me. I never lost consciousness. She knocked off my cap, tore off my wild rag, and tore off and broke my glasses, but my cell phone was still in my shirt pocket with the flap buttoned down. I was able to call Lynne. I told her I was badly hurt and to come down and get me in our four-wheel drive pickup. She said: "Should I call 911?" but I said: "No, just come and get me."

As I said before, there was a foot of snow on the ground and the road was slippery and hazardous, but she was able to get the pickup down there. In the meantime, I had been able to get into the cab of the truck. When she got there, she tried to help me get into the other pickup, but I was unable to move, as it was extremely painful. Obviously, in hindsight she should have called 911 from the house.

Lynne called 911 with my cell phone, but we had no way to direct them to the field. Lynne got hold of Rita March, who drove to the end of the lane and told the ambulance crew how to drive down to the field. Fortunately, they had a four-wheel drive ambulance or they would not have been able to get down there.

After much effort, Amy Stacher, Deputy Chris Baker, Larry Hicks, Ross Smith and several others got me on a back board and were able to load me into the ambulance. They took me to the emergency room at Fairchild Medical Center in Yreka. In the meantime, Ross Smith, who was there to fix a pump, finished feeding the heifers and the horses with the help of our neighbor, John Lyons. Ross also helped Lynne get the pickup out of the snow.

At the emergency room they ran me through a CT scan and x-rays and found I had five broken ribs, severely stretched ligaments and tendons in my sternum and back, two black eyes, a deep cut on my forehead, a bruised hand, and a large hematoma on my right side.

While I was still in the hospital some of my "friends" came to visit me. I perform Cowboy Poetry at benefits and conventions with other local Cowboy Poets, all of whom are quick-witted jokesters. Three of them, Bill Roberts, Butch Jones, and Phill Laubacher showed up at the hospital together. I know they had plotted this out ahead of time. They thought they would cheer me up by making me laugh. If you have ever laughed with a bunch of broken ribs, you know it's not funny.

They would tell jokes; I would laugh, then wince in pain from the broken ribs. Somehow they thought that was humorous. Maybe some day one of them will break some ribs and I will get my revenge. That's what friends are for.

Milk Parlor at the Bryan Ranch

This story my dad and his brother, Al Bryan, told many times. In the 1920's and 30's most of the ranches in Scott Valley had a few, 15 to 25, head of dairy cows. They milked by hand, separated the cream, fed the skim milk to hogs, and sold the cream to the creamery in Fort Jones or Etna.

The Bryan Ranch had about 20 head of dairy cows. My grandfather, Charlie Bryan, and his two sons, Frank and Al, all milked. They had a milk parlor, a long row of wooden stanchions that held the cow's head so she could not back out until she was released. Hay and/or grain was put in each manger in front of the stanchion so that when the cows came into the milk parlor, they went to their place, put their heads into the stanchion to eat, the milker locked the stanchion and then milked the cow. When he was finished milking, he would open the stanchion and the cow would leave.

My grandfather was milking a big Holstein cow. When she put her head in the stanchion he forgot to lock it. He just sat down on his milking stool and started to milk. When the cow finished her feed ration, she stepped back on Granddad's left foot and pivoted on that foot to leave the parlor. Granddad screamed and yelled and swore at her as he punched her in the belly to get her off his foot, all to no avail.

She pivoted 180 degrees standing on his toes, then calmly walked out of the parlor. By this time she had mashed his left toes pretty good. He jumped back and picked up his *right* foot and hopped out of the parlor on the injured left foot, still yelling at the cow.

Of course, Frank and Al thought that was extremely humorous and were laughing so hard they could hardly talk. Granddad failed to see the humor until they were finally able to tell him he had picked up his good foot and was hopping on the injured one.

Moonshine

I don't remember for sure who told me this story. I believe it is true as whoever told me (probably Gene Selby) had all the people's names, the place where it happened, the time, etc. I forget the names.

The incident took place during Prohibition on the Klamath River. I'll call the man "Bob." Bob and his wife lived in a cabin on the Klamath River. They made a living hunting, trading, mining, and doing seasonal work in the mines or on the local ranches. It seems Bob liked a little alcohol on occasion.

He had a small still in his woodshed and was making moonshine. One day when he had a few jars ready to sample, his friend "Bill" came by. Bob's wife would not let him in the house when he was drinking so Bob and Bill sat down on some blocks of wood and had a sample. The sample seemed okay, so, of course, they had some more. It got on into evening and they were hungry. Knowing they could not go into the house, they took some ripe corn from Bob's garden. Just as it was getting dark, a little spike buck jumped the fence into the garden, so they shot the buck. Moonshine, venison, fresh corn for dinner – they were set.

Well, it took them almost a week, but they ate all the corn, all the venison, and drank all the moonshine and they never built a fire. For those of you who read this and are either too young or uninformed to realize the implication of "no fire," you have to understand there was no electricity or gas available. The only way to cook anything was over a fire or on a wood-burning stove with a fire in it.

Old Lady Wey

My grandfather, Charlie Bryan, lived on the ranch in a concrete block house. The concrete had been poured in molds that made the outside of the house look like it was made of stone. Granddad lived there alone after my grandmother died in 1947. At the time of this story, he was 85 or 90 years old. My father and mother and I lived in a smaller house next door.

Some people named Wey moved on to a piece of property east of our timber patch. Their mailbox was on the county road ½ mile across our place from their property. Mrs. Wey, who had been born in Germany, walked through our timber to get their mail. One day, my dad and I were in the timber cutting wood. We saw Mrs. Wey coming across the flat toward us. I could see she was talking but we had chain saws running and we couldn't hear her until we shut off the saws.

She said to my dad: "I know who you are." My father, also knowing who he was, said: "Oh, who am I?" She said: "You live in that big stone house." Dad said: "No, I live in a small house behind the stone house." Old Lady Wey said: "Well then, who lives in the stone house?" Dad replied: "My father lives there." She said: "My Gott, he must be an old man!"

Mrs. Wey was not noted for her tact. Her husband's name was Eugene, but with her heavy German accent, she called him: "Hewgene." Eugene had been a tugboat worker before moving to Scott Valley and had been in a few wrecks over the years. He only had a thumb and little finger on his right hand.

One day after my wife, Lynne, and I were married, the Weys came to our house. Old Lady Wey introduced her husband to Lynne by saying: "Dis is my husband, Hewgene, vat's left of him." We have laughed about our German neighbor, Old Lady Wey, many times over the years.

Old Timer Telling a Hunting Story per Elmer Holmes

"Well, me and old Jim went huntin' over on Moffett Creek last Saturday. No, no, it wasn't old Jim, it was old Pete. Anyway, we got in Pete's Chevy, no, it wasn't Pete's Chevy, it was my Ford. Wait, did I say Moffett Creek? It wasn't Moffett Creek, it was up Long Gulch. Well, it don't matter, anyway.

When we finally got my old Dodge started, ... no, it was the Chevy. I remember because I had been working on the Dodge all day Saturday, so it couldn't have been Saturday, it must have been Sunday. And as I remember, it must have been Jim after all, because Pete works Sundays. That don't matter, the point is when we got to Shell Gulch...yeah, now I remember it must have been Shell Gulch because we drove right past Ike's on the way.

Ike was out plowing, thought that was funny, Ike don't usually work on Sunday, so it must have been Saturday after all. Anyway, that don't matter, Jim had his .06...no, couldn't have been Jim, he ain't got an .06, he shoots an old 30-30 Winchester Model 94...no, wait I think it was a model 92. You know, they made that Model 92 clean up into the 1930's or '40's. Hell of a gun.

So it must have been Pete, guess he took Sunday off. Oh hell, it may have been Tuesday for all I know. Anyway, that don't matter, the point is...oh hell, I forgot the point.

Pumpkin Jar

My grandmother, Nellie Hovenden Bryan, died at 80 years of age in 1947. She always put up lots of produce from her garden. The fruit cellar contained multiple jars of her canned goods.

My grandfather, Charlie Bryan, lived alone in their house after my grandmother died. Occasionally he would open a jar of this fruit cellar hoard, but people brought him food and he made little progress in using up all the stuff my grandmother had canned.

It was my custom to stop at my grandfather's house on my way home from school. The bus stopped at the end of our lane about ¼ of a mile from his house, which was close to my house. I would bring his mail and visit for a while before I went on home.

One day I made my regular stop and was visiting with Granddad in his kitchen. I was probably about 15 or 16 at the time. He had a quart jar of my grandmother's canned pumpkin on the counter and was going to open it. No telling when the pumpkin had been canned, it could have been 15 or more years old. Granddad had the ring off the jar and was trying to pry the lid up. Those old lids had a rubber gasket seal and over time this one apparently had become glued to the top of the jar and was stuck.

Looking back, I fail to see the logic in what he did next, but this is how it happened. He was standing at the sink and I was sitting at the kitchen table 15 feet away. He said: "There must be pressure holding the lid down. I'm going to release the pressure." So he got an ice pick out of the drawer and punched a hole in the lid. Well, there was pressure all right, but it was all inside the jar and the pumpkin was rotten, very rotten, 10 or 15 years of rotten. It came out of that icepick hole like water out of a fire hose nozzle. It coated the window, the walls, the ceiling, and my grandfather's face.

Granddad jumped back, his face, including his glasses, totally covered with rotten, putrid pumpkin. He yelled: "Help, help, I'm blind!" I was yelling at him: "Granddad, take off your glasses! Take off your glasses!" Finally, he heard me, took off his glasses and said: "Thank God, I can see!"

Cleaning up that kitchen was a big job, and even months later you could still smell rotten pumpkin in there. The rest of the old home-canned stuff in the fruit cellar was hauled off.

Rope a Pig
1944 or '45?

When I was around 8 or 10 years old, my dad bought a new catch rope. This was a big deal. He was proud of his new rope and used to tell me to "run by and bawl like a calf," and he would pick up both "hind feet."

This inspired me to want to rope. We had hogs at this time and they ran in the hog pen and a small hog pasture. We fed them on a cement slab that was in a building (granary/barn.) I tried to run one down to rope it, but was unsuccessful. I had to give up on cats and dogs due to my mother's request. Then I decided to borrow Dad's new rope without permission.

Well, I couldn't run down a hog, so I got some grain, made a big loop on the slab, put grain in it, and got in the loft above the slab and called the hogs. When one got in my loop, I jerked the slack and caught a hog.

Of course, I could not hold onto the rope, so the hog ran into the pasture with Dad's new rope. At this point panic set in. I couldn't get any help getting the rope back because I knew I should not have it. I could not run down the pig because he outran me.

Well, the pig eventually got out of the rope when I picked it up. It was no longer new; mud and pig poop were well dragged into it. I washed it the best I could in the ditch, but it still did not look new.

My mother knew what I had done because I had solicited her help while I was chasing the pig (she refused to participate.) I don't remember being formally punished by Dad when he found out. I don't know how he knew; I thought I did a pretty good job of cleaning that rope. Of course that was 60 some years ago, maybe I forgot.

Second Valley Bucks

Second Valley is the second valley between Deep Lake and Hankwright Lake in the Marble Mountains. There is a little cove in Second Valley which lies on the northwest side, maybe halfway from the mouth to the head of the valley, probably closer to the mouth. It is a place that from the top a deer can lie down and see below, and also have an easy exit to the Deep Lake side of the ridge.

If you approach from below, the deer see you and just drop over the ridge and out of sight. Most other approaches are equally hard to get to see bucks without alerting them. But there is a way from the south around a bluffie rock where you can approach them undetected. I knew this approach and had killed bucks there before.

Frank Bryan, Roy Mason, Roy's nephew, Cub, and three of his buddies from L.A., and I had been hunting some other area, Deep Lake or Little Elk. We were camped at First Valley, which was Roy Mason's "winter camp." The hunt had been unsuccessful. We got to the mouth of Second Valley, via the Deep Lake to Second Valley trail that comes around above Mad Woman Camp (not the old trail straight up the Deep Lake side past Quigley Camp and over at the head of Second Valley.)

I said: "Let's check the cove." So my dad rode with me up the valley to a place where I would go on foot to access the cove. Dad led my horse back to the valley mouth. The rest of the party were to watch for any bucks I jumped. I carefully approached a spot where I could see into the cove and stopped. After a few minutes, a buck stood up from behind a log less than 100 yards away. I shot and the buck went down but then stood up again.

I shot again, the buck went down, slid down the steep hill into some brush, stood again, and I shot the third time. This time the buck went down for good. When I approached by way of the log where the buck was when I first shot, there was a dead buck. At this point, I figured I had two. When I got to where the second buck was, there was a third buck. Three shots, three bucks.

My dad and the rest of the party were not very far down the hill, maybe a half-mile or so. I shouted (this was before cell phones or even walkie-talkies) for mules to pack bucks out. I yelled: "Mule, bring three." It was one of those times when you could hear a half mile like you were only ten feet away, and I heard Cub say: "He didn't kill three bucks, did he?" My dad said: "He shot three shots, asked for three mules. I suppose so."

Those guys were sort of mad that I had killed three bucks. Those were the only bucks that were killed on that trip. We had camp meat and when those fellows went home to L.A. they were able to take some meat with them, and glad to have it.

P.S. Remember this hunt took place in the 1960's when there were lots of deer in Siskiyou County, and in those days the annual limit was five.

Sheriff's Posse Ride

In Siskiyou County in the 1940's and 50's there was a "Sheriff's Posse." They helped the County Sheriff when needed; mostly looking for lost people in roadless areas where they could use horses to search. The posse also had a mounted drill team. A big part of the posse was social, a reason to have posse rides, etc.

My dad, Frank Bryan, was asked to join. He accepted the invitation and was to be initiated on one of their rides. To say that there would be a little whiskey consumed on these rides would be an understatement. The ride took place in the summer and they camped at an old logging camp near Bray in the early afternoon. After some tricks on the initiates, two or three men tried to hold my dad down, but he was an ex-wrestler and ended up throwing all of them.

Quite a bit of whiskey was drunk in the early afternoon. One man, having had a little too much, passed out and went to sleep on the ground. There were horses tied around camp. When my dad saw a mare taking a piss, he ran up to her and half filled an empty whiskey bottle with mare's piss.

He went around offering everybody a drink of "whiskey," but of course, they had seen him collect the "whiskey," so turned him down. All saw him, that is, you guessed it, except for the guy passed out on the ground. Dad finally just put the bottle on a tree stump and left it there.

When this man woke up, he saw the bottle and took a big swig. A lot of men saw him drink, so the story quickly spread. He never did live it down. Ever!

I was packing in the Marble Mountains fifty years later when I ran into this guy on the trail. I kind of knew who he was, but was not sure until I said: "I'm Mike Bryan." He said "Frank Bryan's son?" I said: "Yes." He said: "I'm the guy who drank the mare's piss."

Shot in the Air

Sometime in the 1960's or early '70's, my cousin, Jim Holmes, who lived in Klamath Falls, came to Scott Valley to visit. We decided to go coyote hunting up Shell Gulch. We went up the hill behind our neighbors, the Dowlings.

We didn't see any coyotes, but we only hunted for a few hours. Coyote hunts were sometimes (not always) an excuse to get out in the hills and just be there. Even though coyotes were a menace to our sheep, killing a coyote was not necessary to having a good time.

We got back to the truck and Jim started to show me his new shotgun. He had a new 12-gauge 3" mag. The 3" mag was pretty new at that time and I had never shot one. We always used the old 12 gauge 2 ½". The 3" had a lot more power and more recoil, of course. This 3" mag of Jim's was a Winchester Model 12 pump. I had a Winchester Model 12 pump in 20 gauge, so the gun "came up good" for me.

Jim says: "Want to shoot it?" Well, that's like sic 'em to a dog. So he loaded it up, handed it to me, and says: "I'll throw this beer can up in the air for you." This was before aluminum cans, so a beer can had enough weight to it so you could throw it. He tossed the can in the air, a nice high throw, and I missed it clean. Jim picked up the can and says: "Ya missed, here let me throw it again."

Well, I didn't think I should have missed that can. The thought went through my mind that perhaps there wasn't any shot in those shells. Jim loaded his own shells and it would be easy to put in powder and wads with no shot. He had loaded the gun, so I didn't handle the shells to notice the weight difference that no shot would make. I didn't say anything, but when he turned his back to me to throw the can, I laid the shotgun down and picked up my 1903 Springfield 30-06. When the can reached its high point, I shot and hit the can solid. It sailed out in the brush 200 feet.

Jim, of course, knew the difference between the sound of a shotgun and a rifle. He looked at me standing there with the rifle in my hand and said: "Let me see you do that again." At that time I shot a lot and knew the little trick of shooting at a thrown object when it reached the top of its arc. But anyway you cut it that was a lucky shot with a rifle. I declined the offer of another shot, saying I didn't want to spoil my 100% score.

I don't know to this day if that shotgun shell had no shot in it or I just plain missed. Some day I'll ask old Jim about it.

Section 2
Poems

Abe

The calf was sick, his nose was red,
"Best give him a shot or he'll be dead."
Tim filled the gun, stepped out of the truck,
"Here you go, baby, this'll change your luck."

The calf got his shot and let out a bawl,
Now here comes Mama, Tim's in for a fall.
She hit him low before he could run,
"Go get her, Abe, this ain't no fun."

Ole Abe was a fat dog and lazy, too,
But he sure knew now just what to do.

Off the truck ole fat dog rose,
He nailed that cow right on the nose.
"Good boy, Abe, now just let her be,
She thought her calf was being hurt by me.

Let's feed this bunch and get out of here,
For if we stay, I greatly fear,
Mama will be back and cook our stew,
And somebody might catch me talkin' to you."

Always Close the Gate

I failed to close the pasture gate, a breach of the number one rule.
Of course the open gate was found by ole String Bean, a long-legged,
    escape-bent mule.
He headed out with half the cavy following him on a good long trot.
I got ahead of them with the feed truck and turned them into the barn lot.

They didn't go back to the pasture, they started headin' for the hill.
Lynne came out to help me, but they blew by her in her Cadillac Sedan
    de Ville.
They thundered down the ranch lane and out onto the county track.
I chased them in the old truck, just me with the dogs on the back.

I leaped out of the truck to turn 'em back, the transmission was still in gear.
I jumped back in to stop it, but the door slammed on my ear.
The truck was in the fence, the mirror bracket tangled in the wire.
My head felt like the clapper in a cow bell, the side of it was on fire.

Blood was squirting out of my ear, like the fresh swallow fork on a calf.
The horses and mules, of course, they just went thundering on past.
A guy came driving down the road as I got out of the truck with blood
    running down my shirt.
He stops and looks me over, says: "Boy, are you lucky, you could have
    got hurt."

I backed the truck out of the fence, dumped ten bales of hay.
I finally got around them again; I was tired of all this horse play.
I sent the dog to turn 'em back, he heeled a big black mule.
The mule broke the dog's hind leg, I sure felt like a fool.

They all went back to the pasture then, through the open gate.
I followed 'em through to feed 'em, this time I closed the gate.
The vet bill was $600 to patch up my busted canine.
I duct taped the swallow fork in my ear and wrapped my head in baler
    twine.

The notch in my ear is healed up now, you can hardly see the crimp.
The dog is doing pretty good, though he still walks with a limp.
I hope I have learned my lesson about always closing the gate,
But I'll probably leave it open next time I'm running late.

Bink and the Mexican Piss Patrol

"Uno mas cerveza, por favor,"
This was the cry from the floor.
Bink made the call one more time,
The cerveza was free, he was feelin' fine.

He got off the boat, his bladder was full,
The car in the lot was too long a pull.
He stopped behind a wall to take a leak,
Slide hollered: "No, Bink, your bladder's not that weak."

He was behind a car irrigatin' a tire,
When two Mexican cops who wanted to early retire
Saw an opportunity to supplement their cash flow.
Bink's public display was going to cost 300 pesos.

They could tell Bink was no fool,
He wasn't finished, he said: Hey, everything's cool."
They could see that ole Bink sure had the edge,
So they let him walk free with just a pledge:

"Next time you're in Cabo and you gotta pee,
Por favor, Senor, go find a tree."

Black Bear

At the edge of the light, just out of sight,
In the dark as black as coal,
An old sow bear, with jet black hair,
Pulled at the buck on a pole.

Her intent was clear, or she wouldn't have come near
A place where humans camp.
She wanted to eat of the sweet red meat,
Just out of the light of the lamp.

I shone a light into the black of the night
To see her face so clear.
She stood her ground, without a sound,
And ate some more of the deer.

Her teeth were broke and that's no joke,
She'd been around for years.
Her breath smelled bad, she looked so sad,
Her eyes were full of tears.

The buck was mine, with horn so fine,
I'm not about to share.
There's an impasse here I greatly fear,
Me and the old sow bear.

"Get out of there, you mangy bear,
Or I'll shoot you in the head."
She looked at me as if to see
If I meant what I had said.

She jerked off a ham and went on the lam
Across the creek in a flash.
She almost sank, then made the bank,
As the bullets made a splash.

Her escape was complete, and she had the meat,
I didn't have time to mourn.
She had left me some, that's better than none,
After all, I had the horn.

Butch

Ole Butch, he went to Idaho, where the bull elk bugles clear.
He had never hunted bulls before, but he certainly had no fear.
They took a white-eyed Appie on which to pack their goods,
She was experienced at the task of packing in the woods.

This Appie had one good eye, but only good by day.
At night she was so blind she couldn't find the hay.
They pulled her along, a'stumbling in the falling snow,
They had to stop real often, the going sure was slow.

Seeing as to how it was so cold, ole Butch had a thought;
He knew they would not need the snakebite medicine they had brought.
Now he didn't want to pour it out upon the ground,
He knew that was not considered environmentally sound.

So he says to his ole pard, "Let me have that jug,
I'll take a little sip while we're waiting for this plug."
They had a little sip, and then they had some more,
When they got to the cabin, they couldn't find the door.

The cabin where they were staying, it had another guest.
This other border was often called a gol' darn varmint pest.
Now old Butch had some new choppers under his moustache,
He kept them by his bed at night so they wouldn't clash.

Those shiny teeth sitting there, ole varmint could not resist,
He stole them away but left a trade so they would not be missed.
The trade he made, in pack rat terms, surely it was just,
For shining teeth he had traded something great, I trust.

A treasure, just recently acquired, out by the hitching rail,
Where the Appie mare had by chance just lifted up her tail.
A pack rat's delight, of trading value extraordinaire,
A green, glistening gem left in the snow by the Appaloosa mare.

Ole Butch's pard awoke before the dawn to brew the coffee strong,
"If we don't get out and on the trail, the bulls will all be bedded down."
Butch grabbed his teeth and slid 'em in before he was awake,
His mouth tasted of stale booze, but he knew he'd made a mistake.

He touched his palate with tip of tongue,
And gingerly he sniffed the aroma of his thumb.
He knew that smell, there was no doubt,
He spit and gagged and gave a shout.

His poor ole pard was the target of his pitch,
Ole Butch swore his partner had made the switch.
"It wasn't me, I'm as innocent as a newborn babe,
But I heard a rat last night, I'll bet he made the trade."

Then Mr. Rat appeared from behind a stick of wood,
His distorted smile showed more teeth than it should.
Instead of the customary two incisors protruding from his snout,
He had four, plus two canines sticking out.

Ole Butch jerked his iron and lined up on that toothy head.
He drilled those teeth dead center with two hundred grains of lead.
Now there is a little moral to this story true,
Don't shoot off your mouth if you like to chew.

Butch's New Heart

Butch went down to the big city, they said his heart was bad.
The doctor had a new one for him, boy was he ever glad.
They sliced him open and stuffed his new heart in,
His heart is pumping great, it has him lookin' kind of thin.

All of this is well and good, we're sure glad he's fit as pie.
But think about that heart he got, someone had to die.
I tried to call ol' Butch to find out all the facts,
But all I got was his answering machine, "Leave a message, I'll call you back."

Now we can dream of the heart we would have chosen,
Kept fresh and clean and pumping, not decomposin'.
But the choice was made by the San Francisco doctor 300 miles away,
When the call came in "We've got a heart" ol' Butch didn't have a say.

It could have been a thief, shot down in the prime of life,
Or maybe it was an alcoholic bum stabbed by a rusty knife.
How about a pretty ballerina, spinning on her toes,
Or some San Francisco dandy, sniffin' on a rose.

They say that pigs and humans are a match, or mighty near.
You don't suppose Butch got an animal heart, maybe a road kill deer.
Maybe a dog or a bull or a mighty grizzly bear.
Perhaps we can tell when his back starts growing hair.

It's been six months or more, and ol' Butch is feeling grand.
He's doing more and more, and on most anything he'll take a stand.
Now he's eating most of the time and putting on the weight,
He's been exploring caves a lot, maybe he's going to hibernate.

Cowboy Poets

Cowboy poets have been around since the beginning of time,
They're just ordinary cowboys who speak in rhyme.

They're no different than other cowpokes,
It's just that when they tell their jokes they sound different from other folks.

This all may sound kind of funny.
Hell, they sure don't make very much money.

But sometimes at night around the camp fire,
When trying to best the tale of the last liar,

You have to be sure the facts are as true
As the last time you related this tale for the crew.

So it just seems to help the old brain
To remember the story and tell it the same,

If you make the end of every last line
Just the same time after time.

Denny

"How many times have you been to Lake Hogun?"
"After today's trip, that would be one."
He laughed and he joked, he worked hard and had fun.
He loved to go new places, anywhere in the sun.
The note at the corral gate, written in black ink,
Said: "You can ride the App, he bucked Angie in the drink,
Take a left at the fork of the track,
Stay on top of that short hog's back."

He showed up at the camp,
He was leadin' four mules,
He was ridin' the App,
He was teachin' him rules.

We gave him tough packs, the long ones, the steep,
He would go there and back and not lose any sleep.
He would ride any mule or horse in the string,
He laughed at the tough ones, he took the bee stings.

Coming out of Elk Camp, seven mules they led.
Packing camp gear, elk meat, and a big trophy head,
Doug in the lead, then Ira, then Denny.
The trail was narrow, the switchbacks many.

Something happened real fast, we never will know,
A wreck in the making, it never goes slow.
Mules down the mountain, dirt in their hair,
Men on the ground, fear in the air.

Doug off to help, Ira lying in the dirt,
Denny's off the trail, they know he's real hurt.
He died in Doug's arms, so fast it's not real,
His last minute on earth he didn't even feel.

He met death with his spurs on, not ready to leave.
We miss him down here, we surely will grieve.
Up there in heaven is a young horse wearin' his gear,
And Denny's in the saddle, ridin' without fear.

We miss you down here, we mourn and we cry,
We'll see you up there in the sweet by and by.
So set your hats back, boys and tip up the Coors,
Here's to you, Denny, the last can is yours.

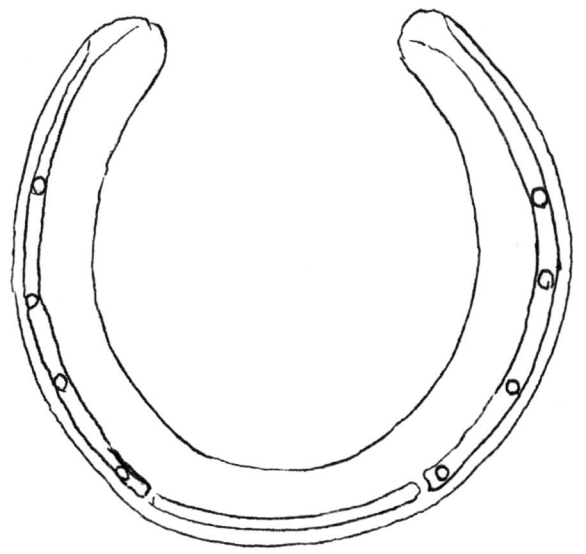

Gus and Call

We were bringing hunters out of Bloody Run; some would say we were fools.
The trail was rough and steep, all of us were riding mules.
By the way, these dudes were hunting bucks with a bow,
One of them had killed a dandy he was bringing home to show.

He was Pope and Young caliber and the velvet was pristine,
The horns were on top of old Call's back so we could keep them clean.
The hunter was riding Gus, a mule of great fame.
Of course, you all know where Gus and Call got their names.

Gus went through some brush and jumped a little stream.
The dude clamped his long legs into Gus' flank and gave him a little steam.
When I came upon the wreck the dude was in the dirt,
Bink was being towed behind Call on his belly, tearing buttons off his shirt.

He was trying to pull old Call around and he was yelling: "Whoa!"
He was "whispering" to him nice and soft, it sure was quite a show.
Finally old Call came around and I saw Bink standing by his neck.
He was telling him he was O.K., that the dude had caused the wreck.

That morning we had ridden past Trinity Lake; saw the water skiers
    flashing by.
Bink sure was impressed, said: "I'd like to give that a try."
So when he was being towed alone behind old Call all fun and fancy free,
He was trying to stand up like he was on a water ski.

How Come You Carry a Gun?

"Why do you carry a gun?; I've never seen anything to shoot,"
Sez the New York lady to the packer who had one in his saddle boot.
Now this ol' packer, his name was Jim, was tired of justifyin' his right.
He'd used that old Winchester for a lot of years, once to stop a fight.

He looked at her hard and he cocked one eye,
He curled his mustache and he let out a sigh.
"Lady," he sez as he cleared his throat and shifted his chew,
"I'll tell you a story 'bout packin' a gun, and I swear it's true."

"I'd packed in early high above the woods,
A long string of mules with supplies and goods.
Came round a ledge covered with shale,
The winter snow had taken out a part of the trail.

My horse stepped through but loosened the ground,
The lead mule went over the edge and a hundred feet down,
She broke both front legs, the bones sticking through,
A big gash in her side and on her head, too.

She was hurt real bad, she knew she was done.
Her eyes pleaded for death, I didn't have a gun.
I couldn't go to her to put her out of her pain,
I had to move on to save the rest of my train.

She suffered, I know, in the hot summer sun,
Suffered unnecessary for want of a gun.
So, lady, you see, that's a reason, just one,
Why this ol' mountain man always packs a gun.

L. A. Flats

These mountains were built from Trinity to Hilt for mountain men to ride;
For mule or horse, this is just the course, don't think that I have lied.

Folks from the South stand with open mouth to see the mountains tall.
They come to stare at the peaks all bare before the snow in fall.

They come in cars from afar, some will come to stay.
These southern lords that come in hoards, mostly from L.A.

They think they're stars, they drive big cars, they're not so very smart.
They wear new hats, we call 'em "flats," they are a breed apart.

They stand in place, this southern race, and stare up at the sight;
They seem possessed in their northern quest, for a place to light.

They would kill what's here with their flatland fear of snakes, and cats, and toads;
For can't you see this land would be peak to peak with roads.

You southern lords that come in hoards, mostly from L.A.
Who won't ride the course on a mountain horse, go someplace else and stay.

Locking the Gate

Protecting the deer from being shot,
Protecting the fish from being caught.
Protecting the bear from pursuing hound,
Protecting the hare from shotgun sound.

Saving the goose from hunter's plate,
Saving the wilderness, locking the gate.
Leave all the trees to grow, then rot,
Don't use a squirrel in the old stew pot.

The gold in the ground is there to stay
'Cuz some preservationists want it that way.
Sometimes I wonder at the laws of the land,
If those that made 'em had a calloused hand.

If those law makers in suit and tie
Had to live from the land, they'd surely die.
There is only one source of real new money,
It comes from trees and grass and meat and honey.

It's not made by people with high pay
It's made by men who don't shave each day,
By women in the fields from sun to sun,
By miners and loggers and cattlemen.

By fishermen and farmers and cat skinners, too,
By men and women like me and you.
This land was made for man to use,
To produce and replenish, not abuse.

Young trees sprout where old were fell,
New grass grows where cattle dwell.
So let the people who know the lands,
Those who work it with calloused hands,

Keep it and use it, and it will give
A life to those who on it live.
Ducks and geese will come to share
The bounty that is everywhere.

For those of you not in the know,
There is a place where you can go,
Where you can spout and wave your hand,
Up here we call it... Disneyland!

Lucy

Lucy was a mule broke to pack and ride,
Dennis bought her at the Bishop sale for his lovely bride.
Brenda rode her all that summer and never had any spills,
In the fall Dennis borrowed her to go hunting in the hills.

He was riding in the lead, the mountain sure was steep.
He came around a bend right under Cabin Peak,
A four-point buck was standing there in the morning sun.
Dennis jumped off Lucy and jacked a shell into his gun.

He squeezed off a shot when the shoulder met the cross hair.
Lucy quit the trail like she'd been boogered by a bear.
There wasn't any way but down and that's the way she went.
She jumped and plunged and scrambled and rolled 'til she was all spent.

She was half a mile below the trail, it seemed more like three,
Standing in the rocks and brush by a very little tree.
Dennis and his brother Doug scouted all the ground,
There was no way out below and they couldn't go around.

They tried again the next day; they didn't have any luck,
And just to add to the deal, Dennis never touched the buck.
They rode out to the trailhead and Dennis called his wife.
Then he called Bink: "Ya gotta help, this could mean my life."

When Brenda heard the news she wasn't one to pick a bone,
She just said: "Dennis, if you can't get Lucy out, don't bother coming home."
They headed into camp loaded down with tools,
They spent three days a'building trail, they worked like bloody fools.

They made a trail up out of there through the brush and rocks,
Lucy wasn't hurt real bad, just some cuts around the hocks.
When Dennis goes a'hunting bucks in the hills next fall,
I think that probably Lucy will be standing in her stall.

Monte

When we left the ranch about quarter to nine,
We packed up light, to make some time.
"Let's pack old Monte and the bay mare,
They sure are an inseparable pair."

His withers were high, he had long legs,
But on his back don't pack any eggs.
His hide was yellow, his heart was black,
But he could travel and carry a pack.

Halfway there, by a big cedar tree,
Something spooked the buckskin, maybe a bee.
He bogged his head and started to buck,
The lash rope busted, just our luck.

The top pack went, it lit in the rocks,
The next time down he jarred loose a box.
When the whole damn outfit was on the trail,
Ole Monte just stopped and switched his tail.

I wasn't mad, never lost my cool,
I resisted the urge to thump that fool.
Now that's a lie and to my shame,
Right then and there I changed his name.

Ole Smoky

He was going to be 24, he didn't seem that old.
Seems like only yesterday that he was foaled.
His old joints were stiff and one eye was turning blue,
And when he walked he wasn't tracking quite true.

His coat was dull and he shed off late in the year.
His hooves were split and he only had one good ear.
I knew it was time to lead him out and put him down,
But as I thought of the good years he'd been around

I would sort of choke up and put it off 'til another day.
Then one morning he didn't come in for his hay,
His teeth were bad, and I knew his mouth was sore,
His time had come, I could put it off no more.

I checked the shells in my worn old saddle gun
And headed to the back pasture toward the rising sun.
As I walked across the frozen grass to do this job so hard,
I thought of all the years ole Smoky had been my pard.

Then I recalled the day ole Smoky took a piece of my arm;
This old pony was starting to lose some of his charm.
And how about that time he jumped sideways just on a whim;
When I went off, I got hung upside down on a tree limb.

Or the time I bragged about his rope horse skills around the campfire,
And he blew up in the roping corral and made me out a liar.
Or the time he broke his picket rope and ran off into the night,
He left me afoot for 20 miles, took the mules too, that ain't right.

I reached up to my nose and felt the crooked bone,
And remembered the time the old buzzard dumped me out there all alone.
Seems like every time I hit the dirt,
Ole crow bait was involved, he never was the one hurt.

As I got closer to the trees where he was standing in his favorite place,
A little smile started to break out from the corners of my face.
As I jacked a shell into that worn old saddle gun,
I thought, "Shooting this old s.o.b. might be kind of fun."

Physical

Went to the doc, hadn't been in a while.
The nurse met me at the door, she had a big smile.
Fill out these forms, there's three pages there.
You'd better have insurance, we want our share.

They told me to strip and get up on the table,
They handed me a cup and said: "Fill it if you're able."
They handed me a night shirt, said: "It ties in the back."
When I got it on it sure left a big crack.

The doc came in reading one of those forms,
Says: "Looks like your old body's been pretty well worn,
Your bones have been broke, your teeth are all loose,
You walk down the hall like a bow-legged goose."

Says: "What do you do that you've been busted all up?"
I says: "I been punching wild cows since I was a pup."
He nods his head like he knew what I meant.
He checks me all over, says: "I guess you're just spent."

I says: "Doc, what the hell is that little rubber glove for?"
He says: Bend over at the waist and look at the floor."

I don't think I'll be going back any time soon,
Even if I live to be old as a coon,
Cuz the last thing he said as he gave me the answer
Was: "Let's check that old prostate to see if you've got cancer."

Rodeo Dance

Whenever I go to the rodeo dance
I like to look at the girls in their very tight pants.
You have to admit they look very nice,
Sometimes from behind I'll even look twice.

Of course, there are some that are a little too tight,
I'll tell you of one that I saw the other night.
She bought her new pants just a year ago,
She only wore them to the dance and the rodeo.

In the year since she bought these tight new jeans,
She had gained a few pounds and broadened her beam.
She pulled and she tugged and she sucked in her gut,
She finally pulled those tight pants up over her butt.

In order to button them, in final despair,
She decided to forego the use of her underwear.
This would have been fine, under normal circumstance,
But not a good idea with very tight pants.

She went to the bar for a drink or a beer,
Had four or five, now she could dance without fear.
She went up to the hall to partake of the dance,
Unaware of the small split in the seat of her pants.

Much to the delight of the cowboy crew,
That little split just grew and grew.
Whenever she dipped they would let out a cheer,
The reason, of course, they could see more of her rear.

She danced 'til the band quit for the night,
Satisfied that she had been a real delight;
For the cowboys kept her twirling in every dance,
Just so they could see the split in her pants.

She went home that night and pulled off her jeans,
Discovered the tear in one of the seams.
Next year it would be my very best guess,
If she goes to the dance, it will be in a dress!

Rolling a Smoke in the Rain

You ever see a cowboy rolling a smoke in the rain?
If you ever did you might think he was a little insane.
Carefully he wipes his fingers 'til they are sort of dry,
Gets a single paper from the packet on the very first try.

All of this is done with his head tipped a little back,
'Cuz he doesn't want that drop of water to drip off his hat,
For if it drops on the paper he has to start again.
It's not an easy thing to roll a smoke in the rain.

He gets the Prince Albert can from his pocket, pops the lid with a click;
Doing this with one hand is really quite a trick.
When the lid made that little sound, his ole pony was sure to hear,
Now he will take this opportunity to pass that slowpoke steer.

He dumps tobacco into the paper, a quick roll to make it tight,
He hopes his matches are not too wet to get a perfect light.
He checks his ole pony to pick up that lazy steer.
Somewhere during all of this he realized his greatest fear.

That painfully rolled smoke of his took on a water drop,
There is no way to salvage a cigarette that's sopped.
He chews it up 'cuz it's all loose.
Next damn time he'll buy some snoose.

Roundup in the Snow

We rode all day with little pay
To get the cattle gathered.
They were bedded now, every last cow,
And our horses were well-lathered.

We camped down low because of the snow,
We knew it was a'brewin'.
We got lots of wood, we knew we should,
Our supper would be a'stewin'.

There was an old stovetop behind a rock
On which to cook our gruel.
A spring was near which we could hear.
The water there was cool.

We laid out our beds to rest our heads,
We didn't have a shelter.
We pulled up tack against our back,
Sort of helter-skelter.

We racked in tight in the dark of the night,
Against the wind a'blowin'.
The wind sang its song, it wasn't too long
'Til it was a'snowin'.

We rolled out at five, 'bout half alive,
Brewed a pot of muck.
The horses were fed, we rolled our bed,
Tried to stir up some chuck.

My hands, they were a'shakin' as I cut thick slabs of bacon
Into a red hot fryin' pan.
The grease was a'poppin', with snow flakes a'droppin'
Slow, as they sometimes can.

Our saddles were on before the dawn,
The horses standin' humped.
Best take some pains and pull up the reins,
Or you'll damn sure be dumped.

The wind had stopped, but the snow still dropped,
Two feet or more by now.
"Let's get a'goin', the way it's snowin'
We'll never find a cow."

Up the line 'til we cut fresh sign
Of cattle on the run.
It was my hunch the whole damn bunch
Went back where they had come.

We rode downhill, Jake took a bad spill,
The leaders finally stopped.
We went around on snowy ground,
Our horses damn near dropped.

Back to the top with many a stop,
There's no way out below.
The cows were beat, they had sore feet,
The goin' sure was slow.

During the night, the cows had taken flight
Because of an ole she-bear.
She was huntin' meat before her long sleep
In her winter lair.

She had killed a calf and eaten half
Before ole mother cow bawled,
That high-pitched bawl that tells 'em all
Danger sure has called.

They smelled fresh blood in the frozen mud
Where ole sow had made her kill.
They were on the run before the first sun,
A bunch with a single will.

We had 'em back and on the track,
Past the bloody calf.
Over ridge crest we had done our best
To execute our craft.

Two of us went back to the track

Of that ole calf-killin' sow.
This wasn't her first, some years had been worse,
We knew that now.

She had killed cattle here for at least five years,
We resolved we'd change her ways.
The damn ole brute we swore we'd shoot,
And end her killin' days.

Our ole cow dogs were jumpin' logs,
Our horses we had to tie.
They were getting' near that ole she-bear
Their barkin' was on high.

Our dogs bayed "treed," they'd filled our need,
The she-bear was cornered now.
In the cold setting sun, justice was done,
She'd not kill another cow.

When Winchesters spoke, they told not a joke,
But a sentence of hot lead.
The bear hit the ground without a sound,
She sure as hell was dead.

This talk is true and I remember it anew
Although it happened long ago,
When I come through the door and see on the floor,
The hide of my ancient foe.

Skipper

We live in modern times, our roads are nicely paved;
Our cars we can drive fast, look at the time we saved.
But if you are ridin' a horse with slick iron shoes
And a dog, he comes a'yappin', you are sure to lose.

You could break an arm or a leg or worse,
They could come and haul you off in a fancy hearse.

So next time you are ridin' down a modern road,
Make sure your ole six-gun has got a plumb-full load
When you see that dog a'comin' and before you hit the ditch,
Pull out that damned ole six-gun and shoot the son of a bitch.

Written on the occasion of Skip Hanna breaking his leg, October, 1991.

Tax Poem

I'm a packer you see, I take folks to the hills.
There are expenses you know, I get lots of bills.
There are taxes to pay on top of that.
I had to pay tax on my new hat.

The tax assessor came out today.
He said there are more taxes I have to pay.
I have to pay tax on my old horse trailer,
And there's an added tax on the new hay baler.

A tax, of course, on your new mule shoes,
And a double tax on that jug of booze.
They added more tax on diesel fuel,
And taxes for your watering pool.

They tax the pasture where I keep my critters,
There are so darn many taxes it gives me the jitters.
I bought some beer for my clients to drink,
There's more tax on beer than you might think.

They tax the oats I feed my horse,
They tax that block of salt, of course.
And if there's anything left at the year's end,
On April 15th they tax it again.

I'm all fed up, I'm about to cry,
But I know in the end they'll tax me when I die.

Valentine's Day Gift

I tried to write a romantic card,
I really, really tried very hard.
"Roses are red and violets are blue,
Honey is sweet and so are you."

Seems like that rhyme has been used before,
So I will try to write something more.
I was going to town to buy some flowers,
But the truck wouldn't start and the prediction was showers.

So I went to the shop to make something you'd treasure,
I had it all welded and sanded and measured.
Then suddenly it came to me like a bolt from the blue,
You already had lots of my art made from horseshoes.

I thought and I studied and I blew all my time.
Guess you will have to settle for this bad rhyme.
I love you, I love you, I give you a kiss,
I know my great horseshoe art you will not miss.

Be my Valentine!

Love,
Mike